The Good and the Beautiful

My Second
Nature Reader

Written by Jenny Phillips

Cover illustration by Shannon Vogus
© 2022 The Good and the Beautiful, LLC
goodandbeautiful.com

TABLE OF CONTENTS

1. **The Skunk**—Sight Words: Group 1 1
2. **Bark**—AR; ALK and OLD 11
3. **Bats**—Sneaky E........................ 21
4. **Storks**—EA; OR 33
5. **In the Sky**—OO Sound 1................ 43
6. **The Pretty Brook**—OO Sound 2......... 53
7. **My Little Sprouts**—OU and OW 61
8. **Elf Owl**—Sight Words: Group 2; INK, ANK, IND 69
9. **Mouse House**—Ending Es That Are Not Sneaky; Words Ending in Y............. 83
10. **Otters**—ER; IR; UR.................... 91
11. **Sperm Whale**—AI; WH 99
12. **Little Brown Wren**—WR; Sight Words: Group 3........................... 109
13. **Red-Tailed Hawk**—AW and AU; Words with Open Syllables 119
14. **Spruce Trees**—Soft E; Consonant + LE ... 127
15. **Spider Wasps**—A Says /UH/ and /AH/ 137
16. **Dew**—EW; IGH 147
17. **Roots**—OY and OI; Contractions 155
18. **Gems**—Soft C and G; Sight Words: Group 4 163
19. **The Shetland Pony**—OA and OE; OW Can Make the Long Sound O 171
20. **The Weather**—Other Sounds of EA; Words with Open Syllables 181

BOOK 1

THE SKUNK

Sight Words: Group 1

The Good and the Beautiful

Illustrated by Valentina Fedorova
© 2022 The Good and the Beautiful, LLC
goodandbeautiful.com

I know where a skunk plays.

It is pretty.

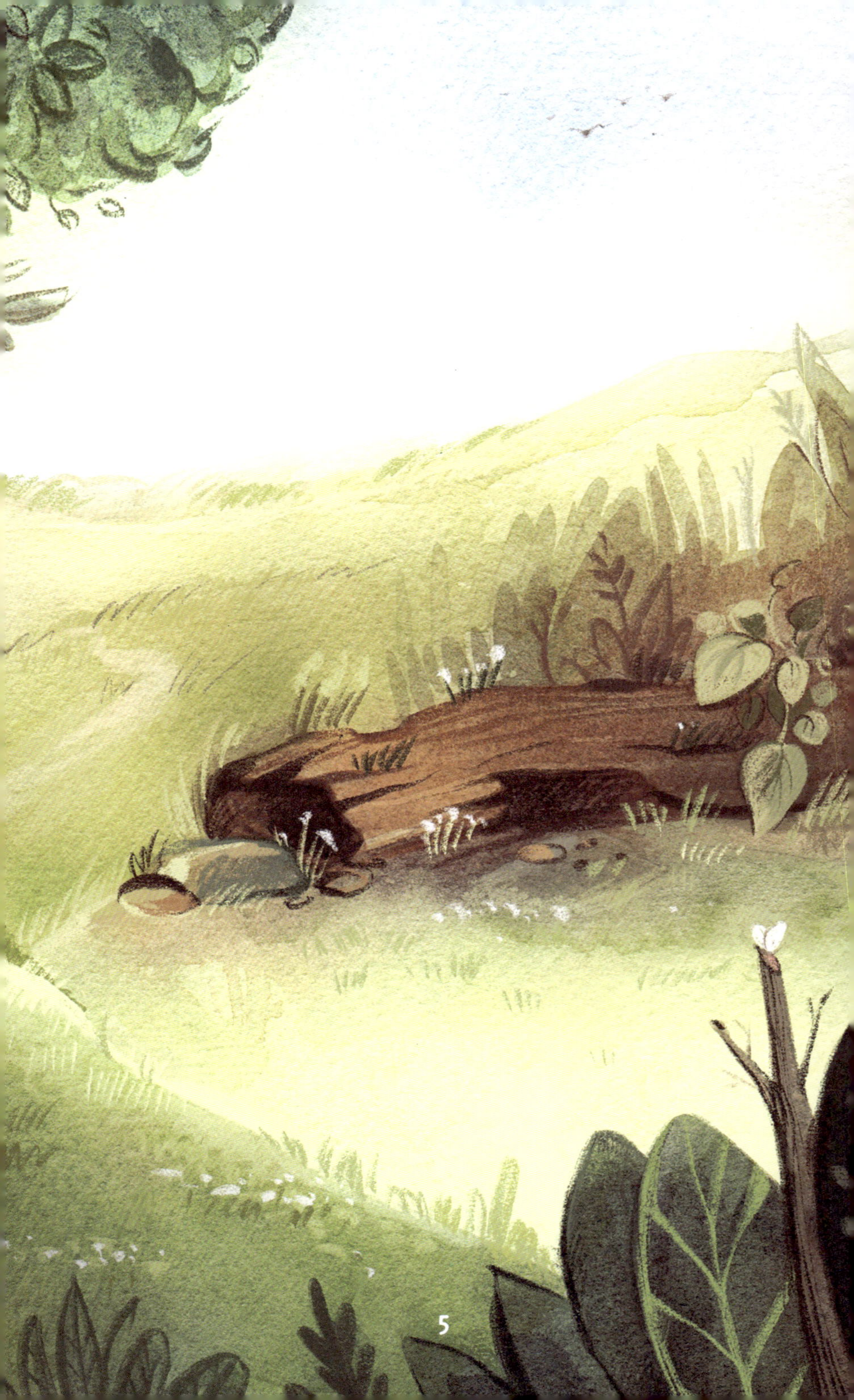

But if I got by it, it would spray me with a very bad smell.

7

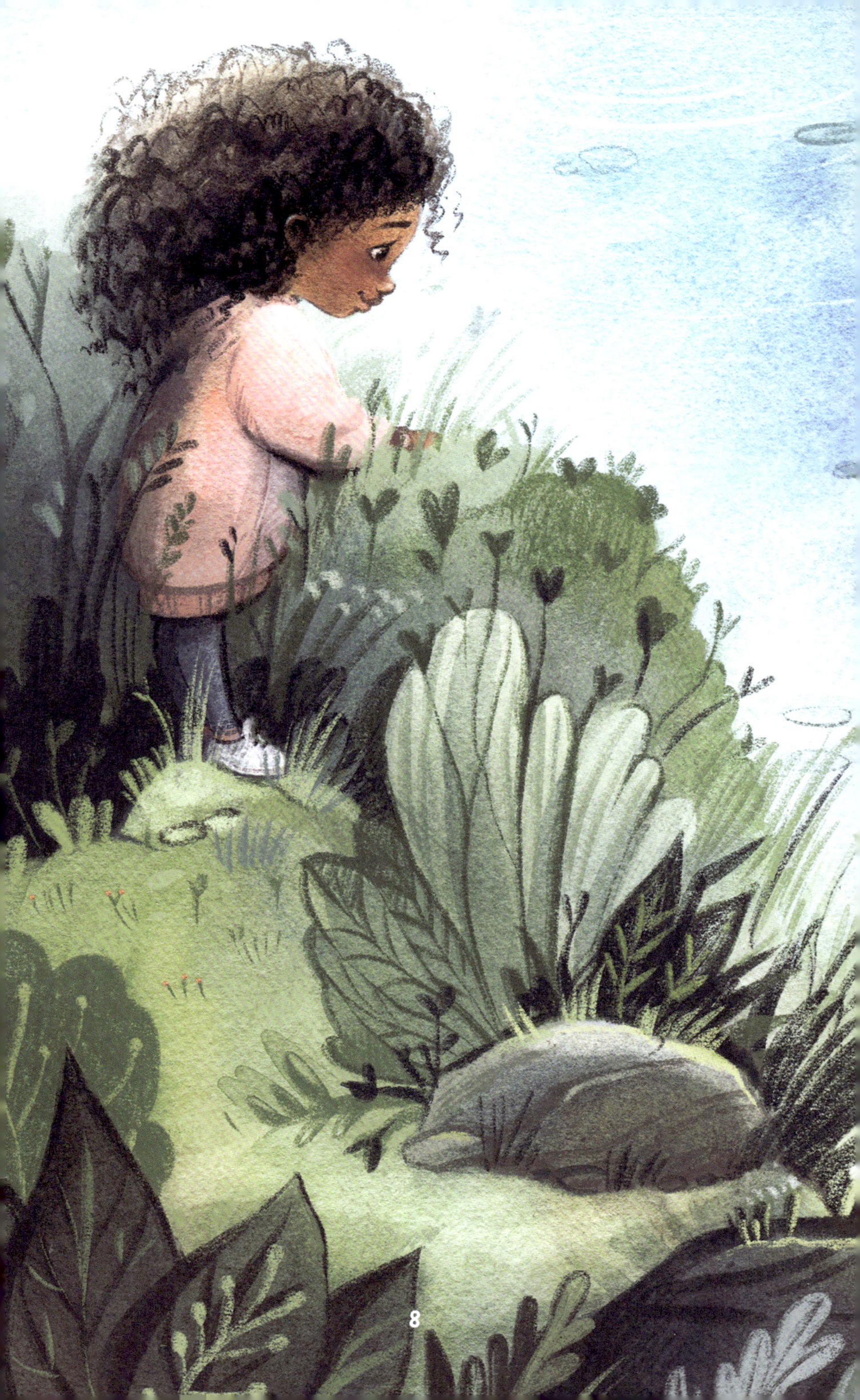

It runs around the blue pond, and I know I should stay far away.

Look at the eight little skunks!

BOOK 2

BARK

AR; ALK and OLD

The Good and the Beautiful

Illustrated by McKenzie West
© 2022 The Good and the Beautiful, LLC
goodandbeautiful.com

I look at the bark in my yard.
I love bark.

Some bark is hard.

Some bark is soft and can peel.

Look at the bark on this tall tree. It is red.

There are black marks on this bark.

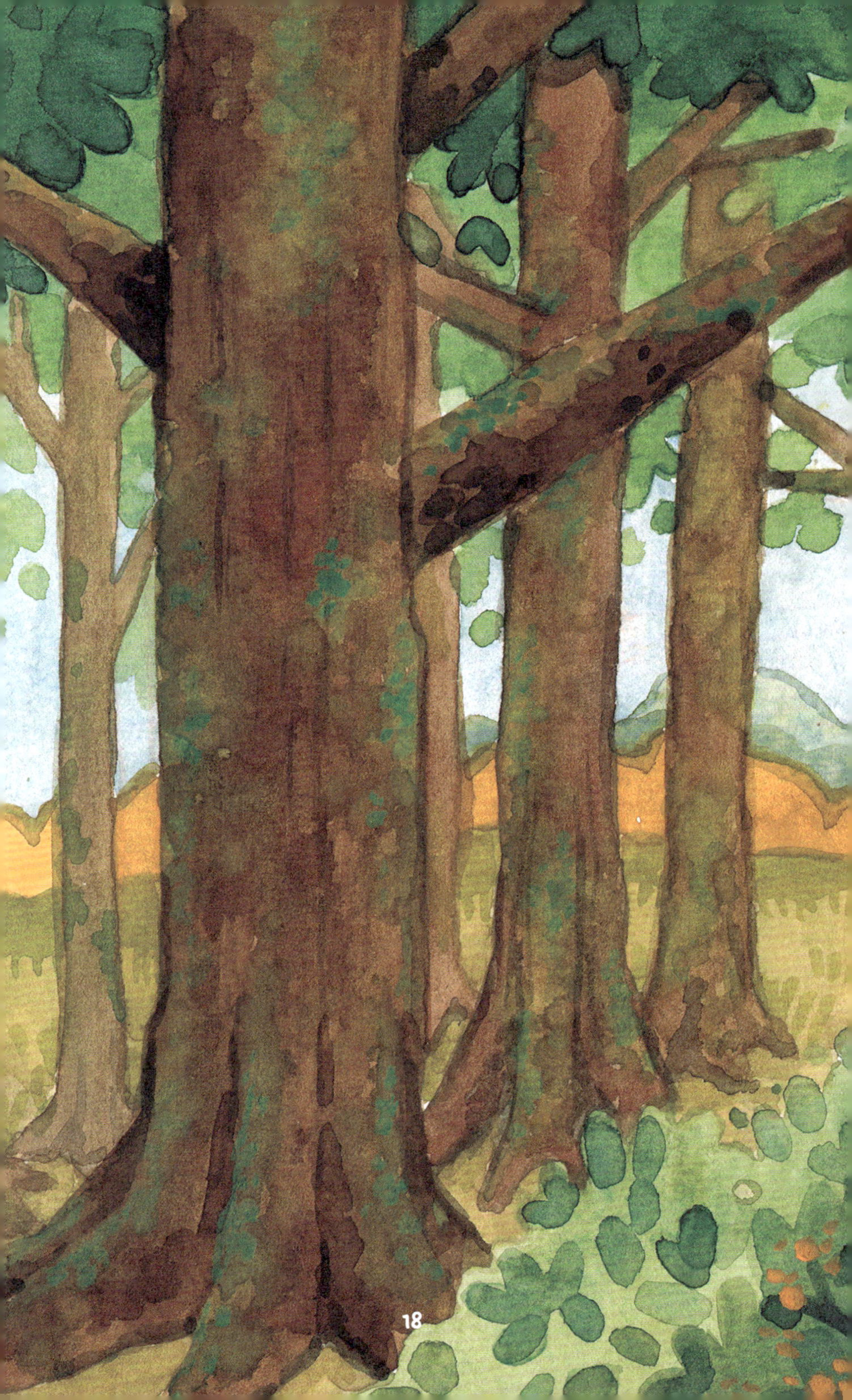

Oh! Look at the moss on the tree bark.

My friend has a farm. I love the bark on the trees.

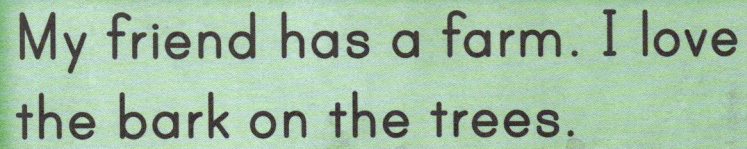

Book 3

BATS

Sneaky E

Good AND Beautiful

Illustrated by Joy Laforme
© 2022 The Good and the Beautiful, LLC
goodandbeautiful.com

Some bats stay in dark caves.

Some bats stay in tree holes.

All bats sleep as the sun shines.

In the dark they wake and fly.

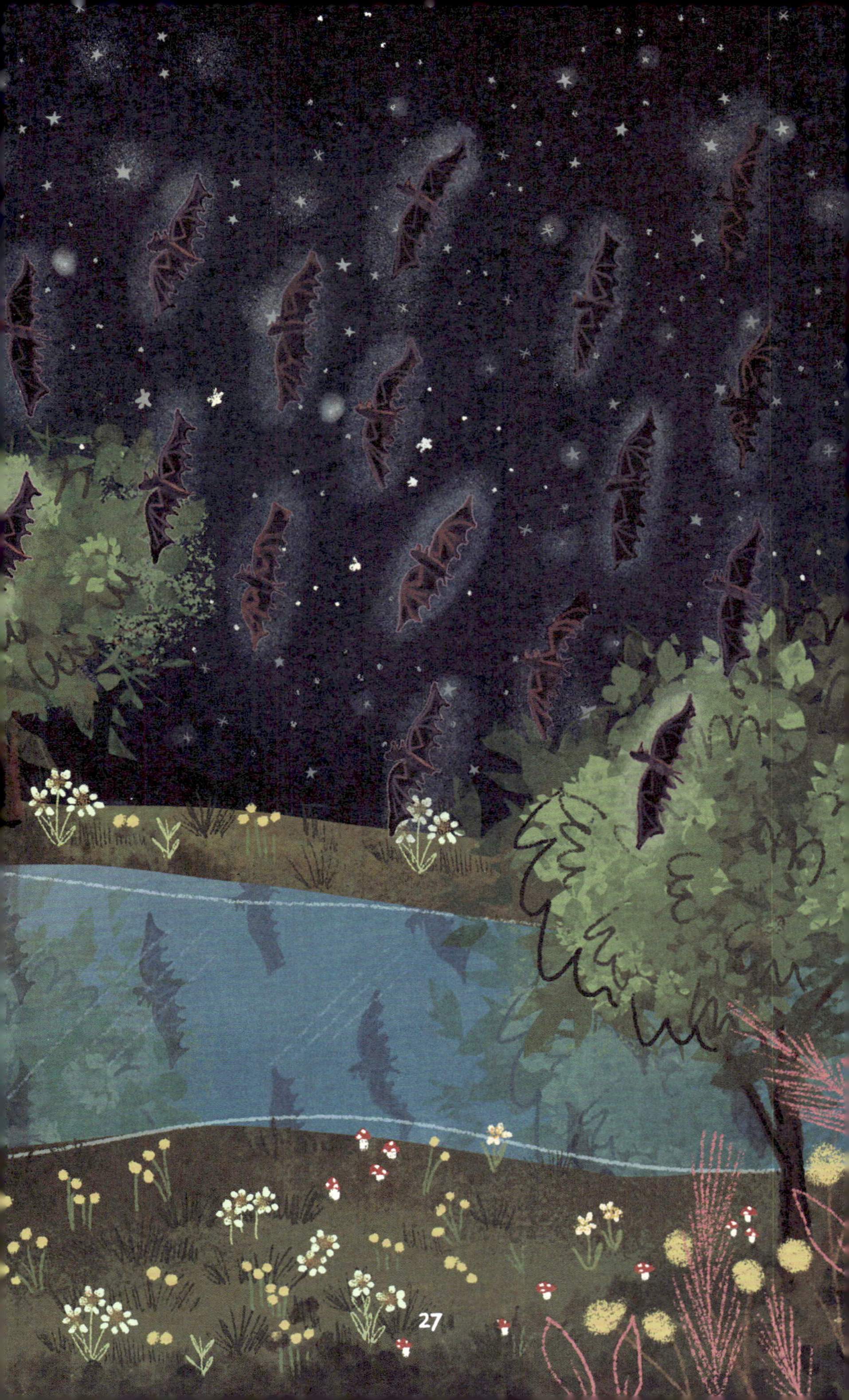

This one chases bugs.

This one tastes a ripe plum.

Not all bats are the same size.

What bat do you like most?

It is late, and they will fly home.

BOOK 4

STORKS

EA; OR

Illustrated by Alissa Empey
© 2022 The Good and the Beautiful, LLC
goodandbeautiful.com

Storks love to be near wet spots.

Some storks make nests in trees or on top of homes.

They use sticks for the nests.

With its long beak, it eats frogs or fish.

39

This stork has a home near the sea.

It is so pretty as it is flying by the beach.

Storks are so neat!

BOOK 5

IN THE SKY

OO Sound 1

The Good and the Beautiful

Illustrated by Larissa Sharina

© 2022 The Good and the Beautiful, LLC

goodandbeautiful.com

The moon shines in the dark sky.

Bats zoom around in the cool breeze.

We hear a *hoot, hoot*.

Soon, the sun rises and shines on the roof and the

pool and the leaves on the trees.

The wind scoots soft ships across the blue sky.

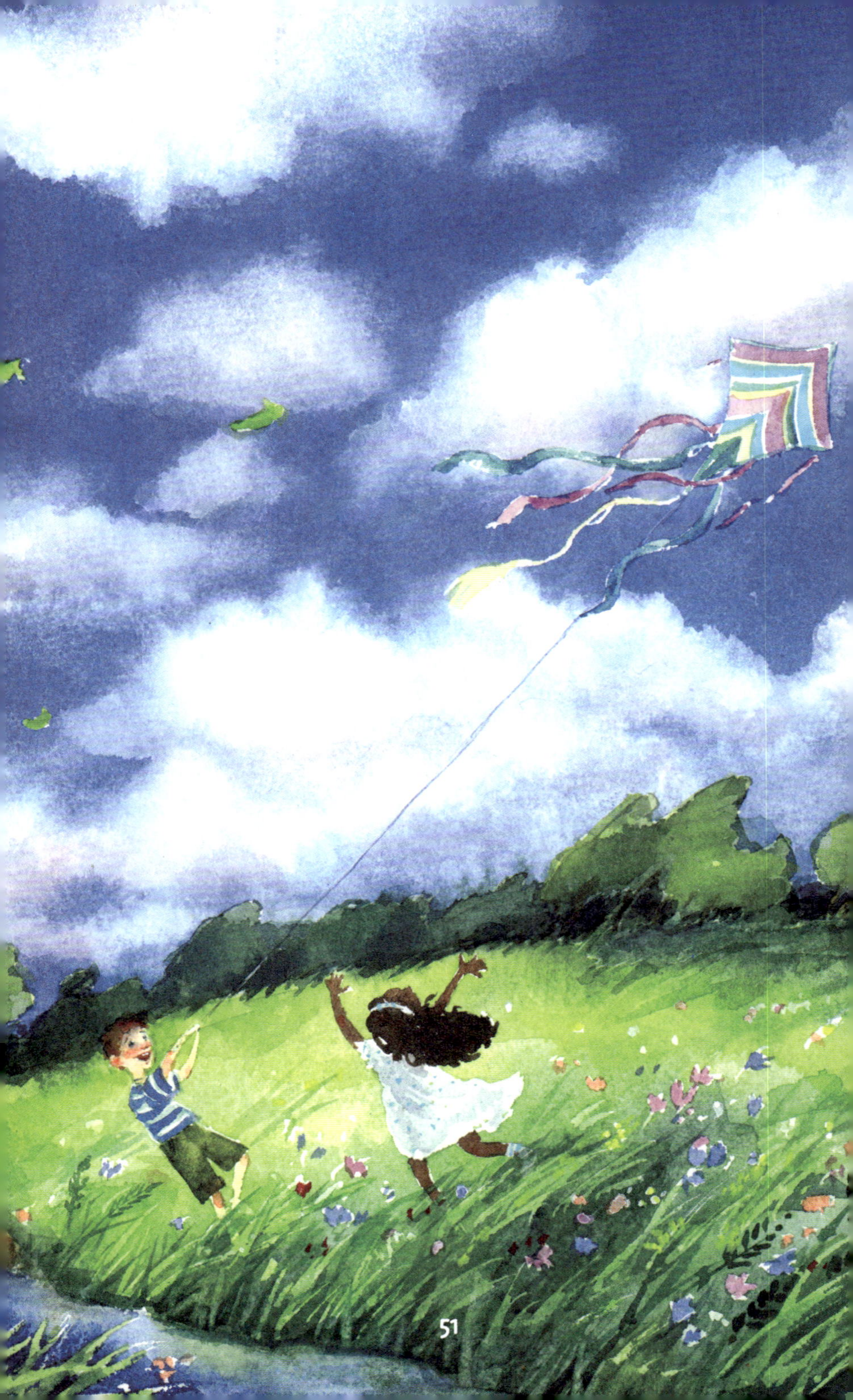

A storm rushes in. The sky is not the same each day.

BOOK 6

THE PRETTY BROOK

OO Sound 2

Illustrated by Raquel Martin
© 2022 The Good and the Beautiful, LLC
goodandbeautiful.com

I stood in the woods by a pretty brook.

I took a good look.

I could spy many neat things.

Sitting on some soft moss,

From th...

fish

I got my book and took notes.

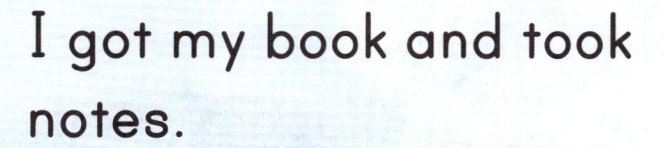

The wind shook the trees, so I held on to my hood and left the pretty brook.

BOOK 7

MY LITTLE SPROUTS

OU and OW

The Good and the Beautiful

Good and Beautiful

Illustrated by Valentina Fedorova
© 2022 The Good and the Beautiful, LLC
goodandbeautiful.com

These are my little sprouts.

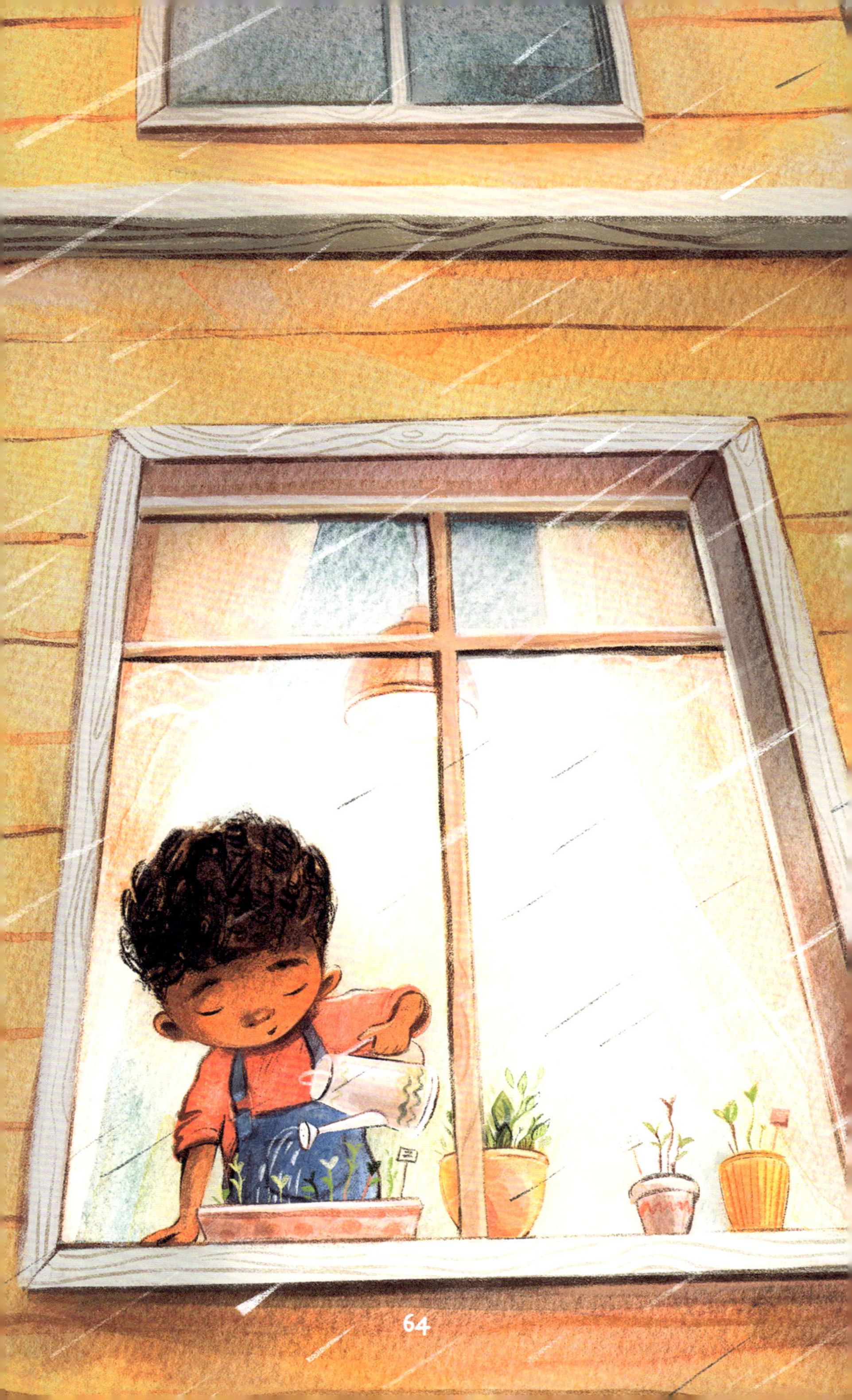

I keep them safe from the storm. I use my spout to care for them.

Now the ground is not cold.

I get down and put my sprouts in brown mounds.

Wow! I am proud of my sprouts.

BOOK 8

ELF OWL

Sight Words: Group 2;
INK, ANK, IND

The Good and the Beautiful

Illustrated by Bojana Stojanovic
© 2022 The Good and the Beautiful, LLC
goodandbeautiful.com

My aunt and I were on a trip.
We stopped in a dry spot.

We went on a hike.

The yellow sun was hot, so we drank a lot of water.

74

"Wow! It does get hot here," I think.

What did we find? Some pink blooms, waving their

petals, and a big cactus.

An elf owl was in a hole.

It was so cute, but it went away.

In the hole were three owlets peeping for their mom.

She came back again.

I thank my kind aunt
for bringing me here.

BOOK 9

MOUSE HOUSE

Ending Es That Are Not Sneaky; Words Ending in Y

The Good and the Beautiful

Illustrated by Alissa Empey
© 2022 The Good and the Beautiful, LLC
goodandbeautiful.com

This mouse finds a lovely house above ground.

The happy mouse moves into

an old nest in the olive tree.

This mouse will choose to dig down into the ground.

It is done making its home.

This mouse lives in the same spot as the goose and the horse: the jolly barn!

OTTERS

Book 10

ER; IR; UR

The Good and the Beautiful

Illustrated by Farah Shah
© 2022 The Good and the Beautiful, LLC
goodandbeautiful.com

Otters have very thick fur.

Webbed feet help them swim better in the water.

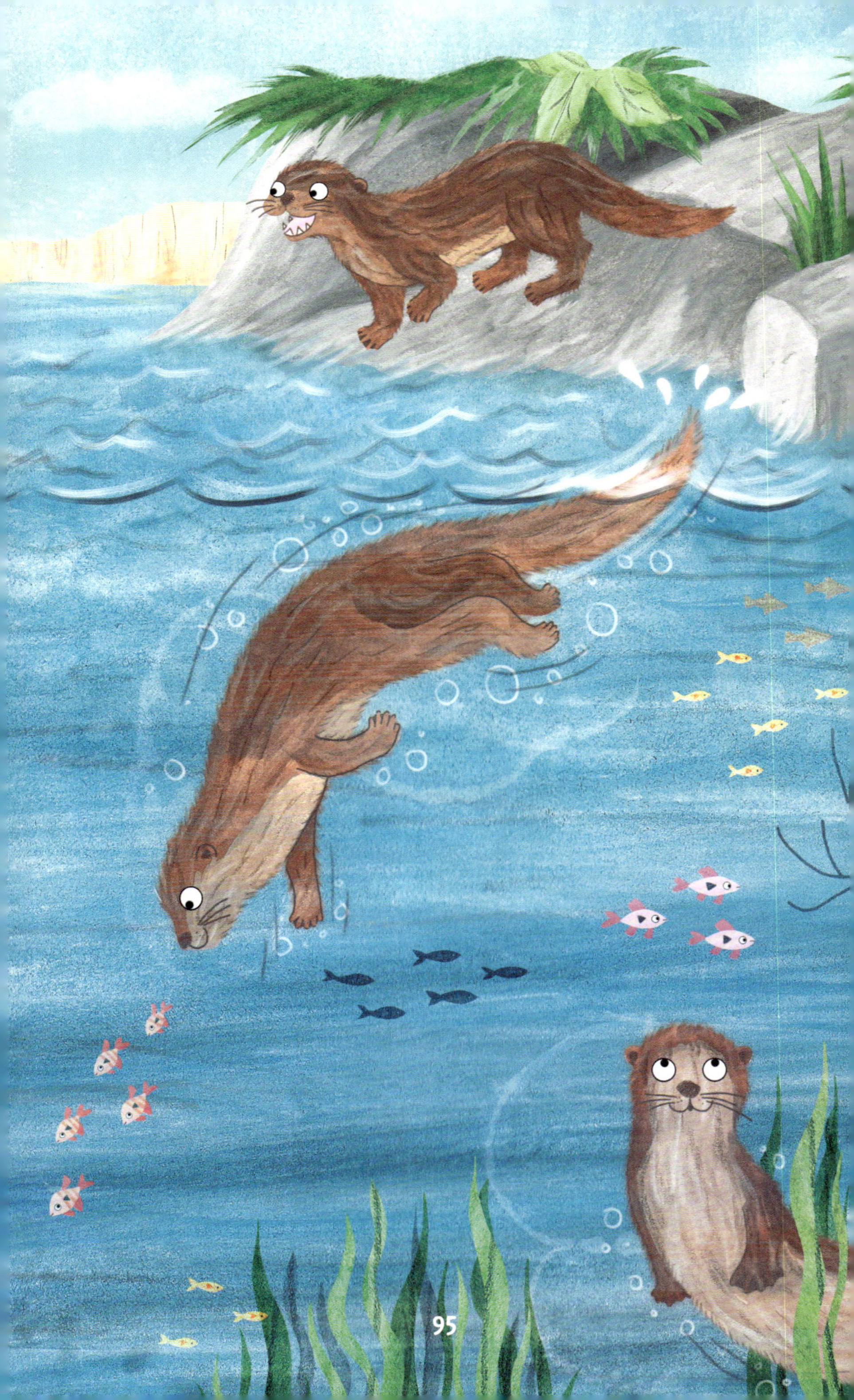

On land they can run faster than you can.

97

They curl up in sturdy dens under the ground.

BOOK 11

SPERM WHALE

AI; WH

The Good and the Beautiful

Illustrated by Farah Shah

© 2022 The Good and the Beautiful, LLC

goodandbeautiful.com

Sperm whales have the biggest brains on the planet.

When they dive, they raise their tails in the air.

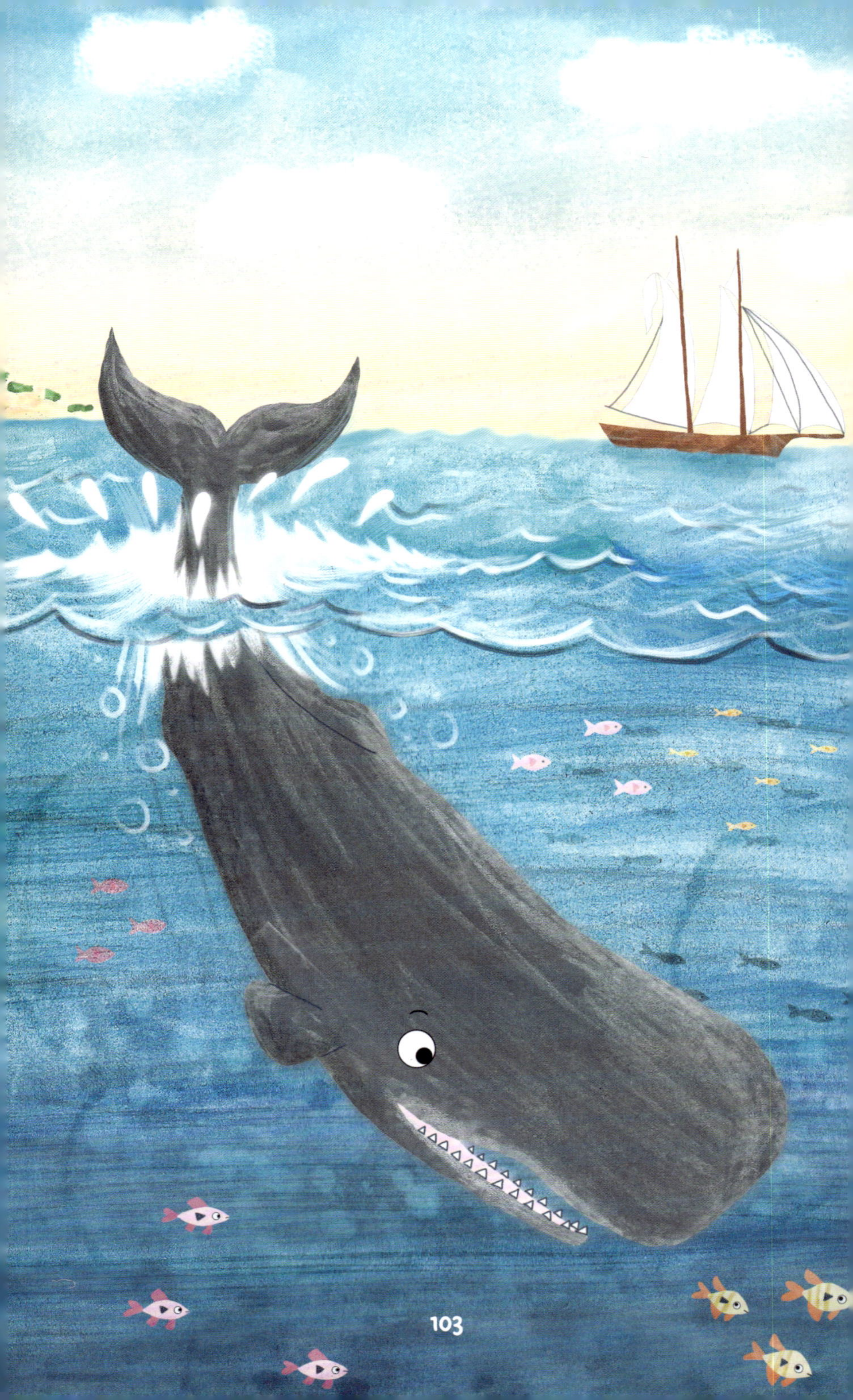

Here is a neat fact: sperm whales nap with their tails down.

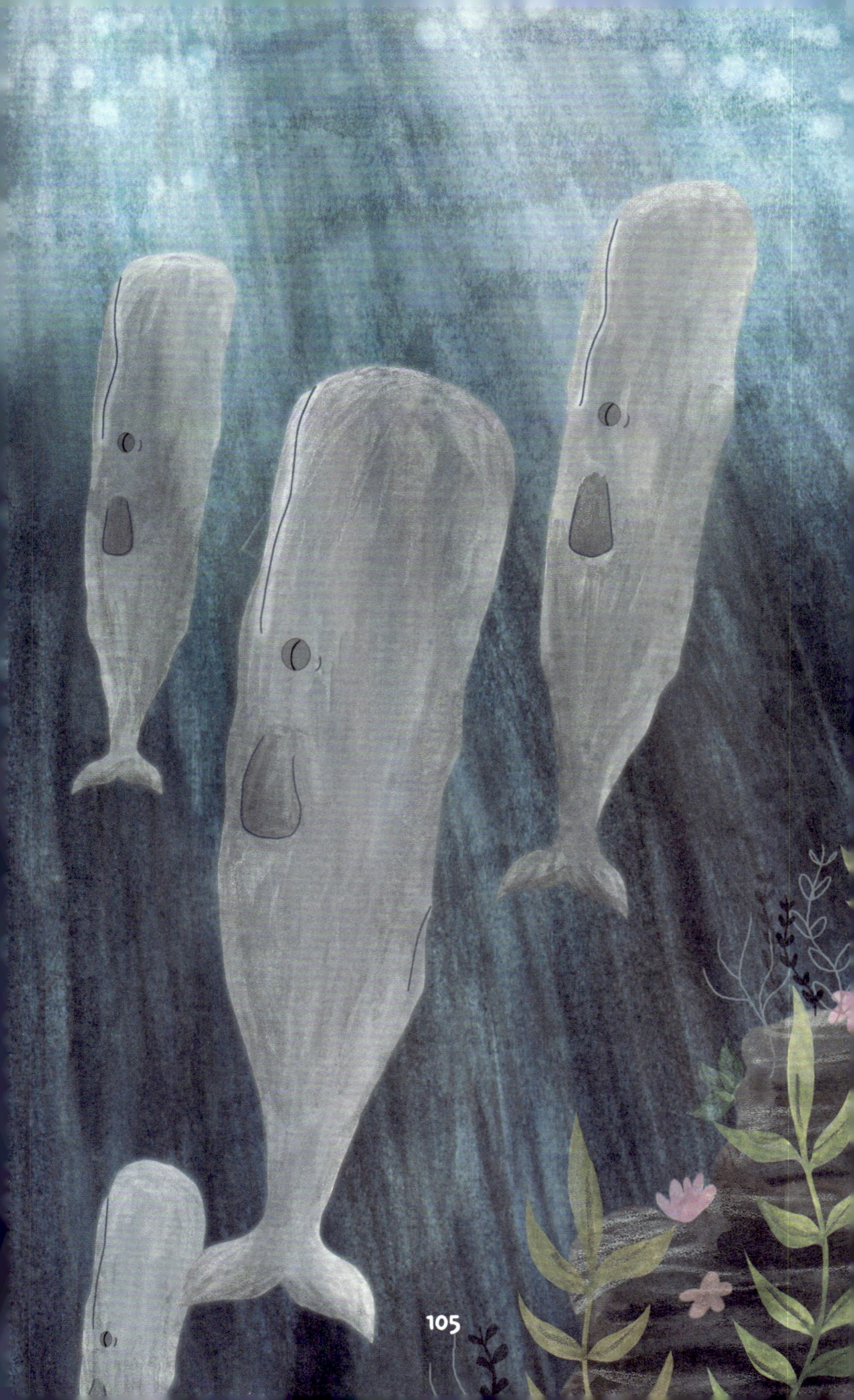

They eat many fish daily,

and they eat sharks too.

I spot a whale from the wharf. It is a whole lot of fun!

BOOK 12

LITTLE BROWN WREN

WR; Sight Words: Group 3

The Good and the Beautiful

Illustrated by Simone Fumagalli
© 2022 The Good and the Beautiful, LLC
goodandbeautiful.com

I love wrens and search for them each morning.

Once I found a hurt wren that had been lying upon the

cold ground. My brother and I kept it warm with a cloth.

We set it in a crate. It sat still and did not wriggle.

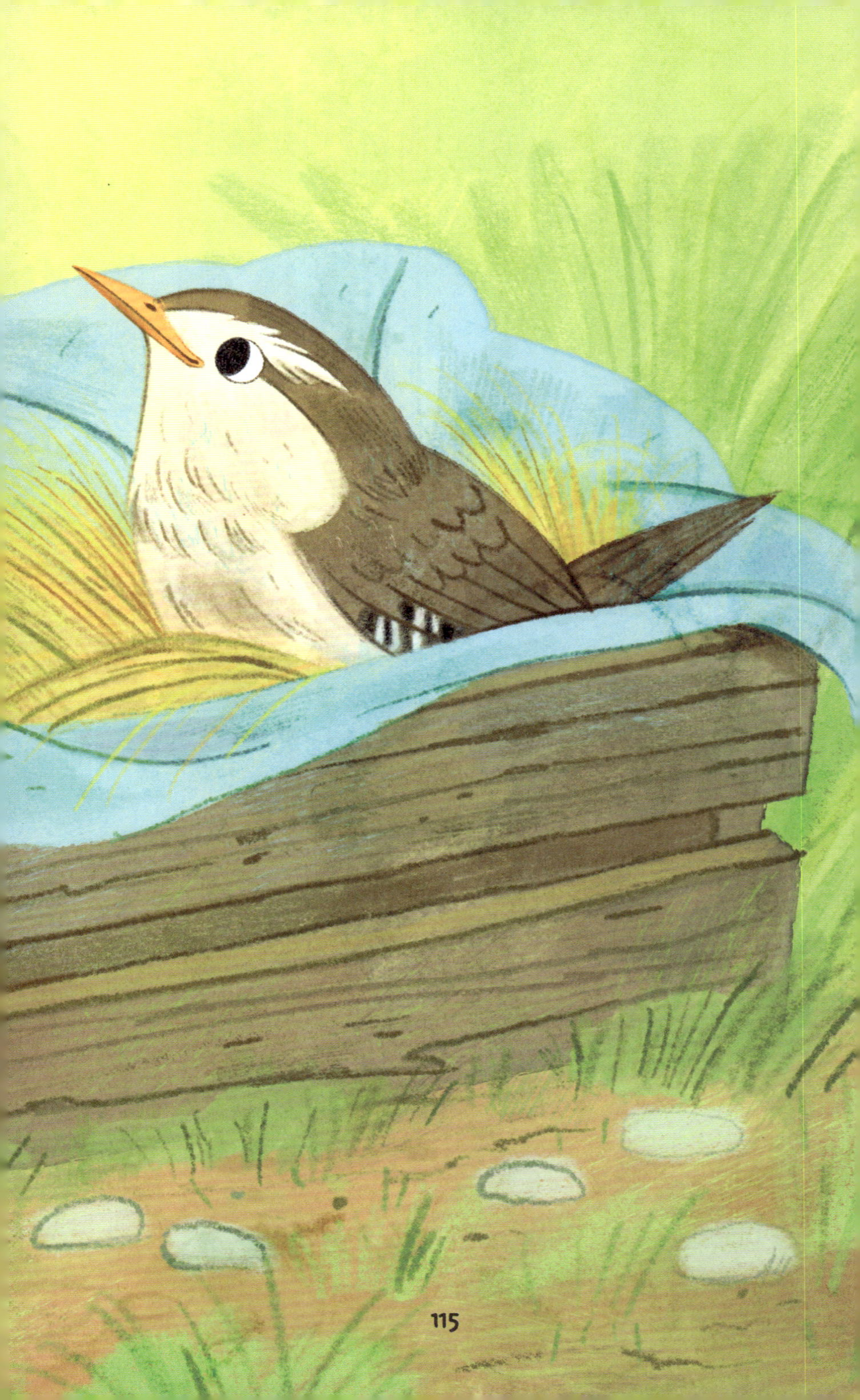

After only one hour, it felt better and left.

Up on a branch with other wrens, it gave us a loud song.

My heart was happy because we had helped it.

BOOK 13

RED-TAILED HAWK

AW and AU; Words with Open Syllables

The Good and the Beautiful

The Good and the Beautiful

Illustrated by Ekaterina Kolesnikova
© 2022 The Good and the Beautiful, LLC
goodandbeautiful.com

It is dawn. A red-tailed hawk launches into the sky.

The hawk was born in May.
It is August, and she is
no longer a baby hawk.

She started flying at six weeks old.

Does the hawk eat acorns?
No, it does not eat any plants.

It eats things like spiders, locusts, frogs, and even vipers.

Two years later, the female hawk holds twigs in her claws. She is going to make a nest!

BOOK 14

SPRUCE TREES

Soft E; Consonant + LE

The Good and the Beautiful

Good and the Beautiful

Illustrated by Natalia Grebtsova
© 2022 The Good and the Beautiful, LLC
goodandbeautiful.com

Leaves on a maple tree change color and fall off each year.

Bird's nest

Maple tree

Maple seeds

A spruce tree stays green all year. There are over 35 kinds

This is a dwarf tree, and it has a flat top.

Bird's Nest Spruce

Colorado Blue Spruce

Siberian Spruce

Can live for 400 years

Red Spruce

Japanese Bush Spruce

of spruce trees. They all have leaves like needles.

Serbian Spruce

Dragon Spruce

Purple Cone Spruce

Sitka Spruce

The tallest spruce tree can be up to 200 feet tall.

Needles smell bad when crushed.

White Spruce

Spruce trees are large. They range from 60 to 200 feet tall.

Blue spruce grows 12 to 24 inches per year. So it will require 30 to 60 years for a blue spruce to grow from a seed to 60 feet tall.

They all have cones. Seeds are in the cones.

Found only in China

Purple Spruce Cone

Engelmann Spruce Cone

Norway Spruce Cone

Seeds

Spruce beetles are a huge problem since they snuggle under the bark and nibble the tree. They can damage or kill a spruce.

Spruce trees are used to make these nice things!

Violin

Harp

Guitar

Piano

BOOK 15

SPIDER WASPS

A Says /UH/ and /AH/

Note: Teach the child that EAR can say /air/ like in BEAR.

Illustrated by Larissa Sharina
© 2022 The Good and the Beautiful, LLC
goodandbeautiful.com

Spider wasps live across the world, even in Africa and the United States.

They wander about looking for spiders. After stinging a spider with a strong venom, the wasp drags the spider across the ground to its nest.

The nest is made among dirt, rocks, or rotting wood.

The wasp lays an egg on the spider.

The young wasp eats the spider when it comes out of the egg.

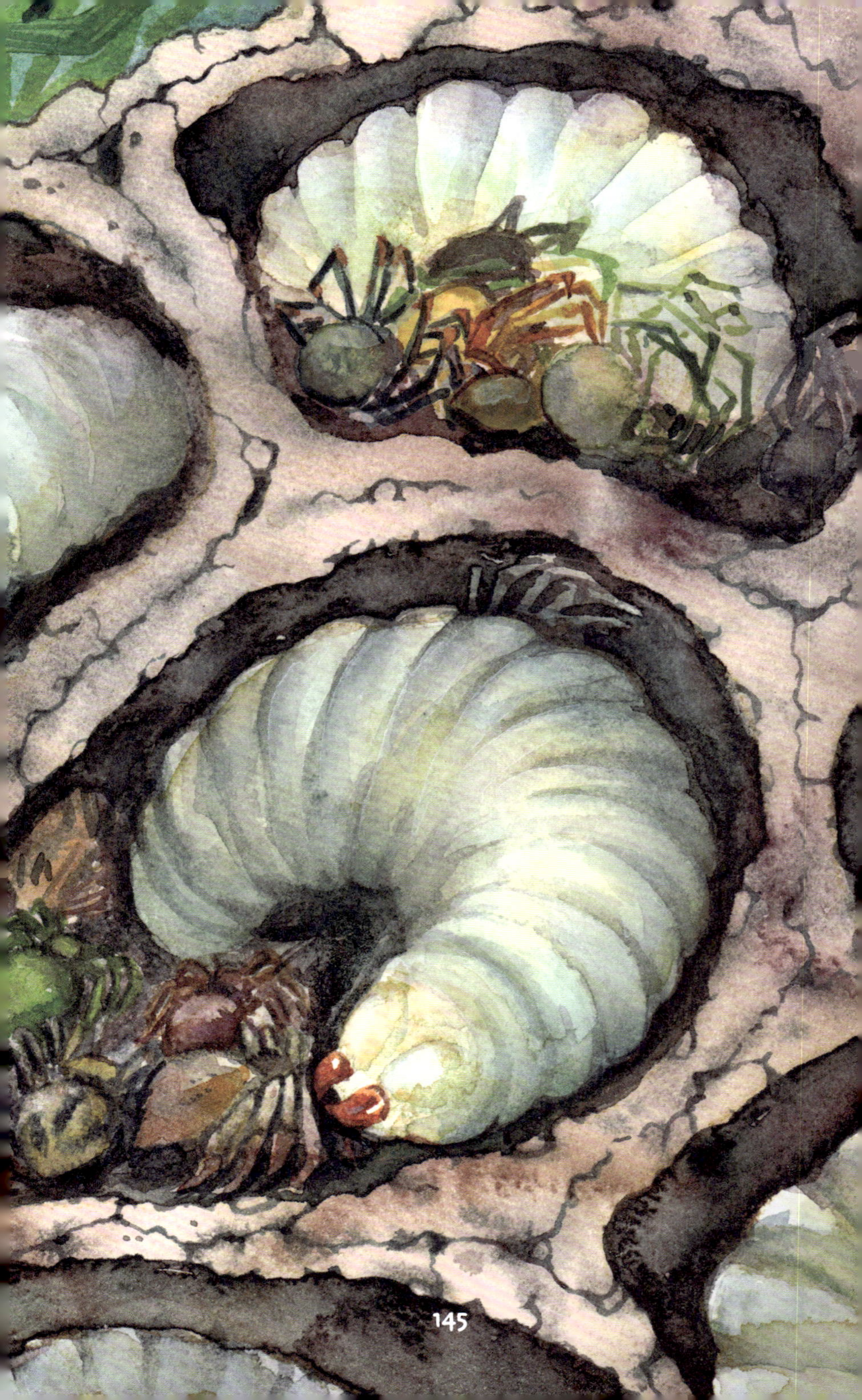

Adult spider wasps do not want spiders to eat—they want nectar from plants.

DEW

BOOK 16

EW; IGH

The Good and the Beautiful

The Good and the Beautiful

Illustrated by Laivi Poder
© 2022 The Good and the Beautiful, LLC
goodandbeautiful.com

A new day dawned. Dew dotted the grass and plants and shone in the light.

Dew forms during calm, clear nights as the air cools

around things. Humid places are more likely to have dew.

I love dew. I looked at the dew that sparkled like bright jewels as a bird flew over the grass.

What a lovely view!

If it is too cold, dew turns into a frost—a lovely sight too!

ROOTS

BOOK 17

OY and OI; Contractions

The Good and the Beautiful

Illustrated by Bojana Stojanovic
© 2022 The Good and the Beautiful, LLC
goodandbeautiful.com

Joy sees the tree stand tall like a royal tower, but she doesn't see the part of the tree beneath the ground: roots.

Some tree roots join together like with these aspen trees.

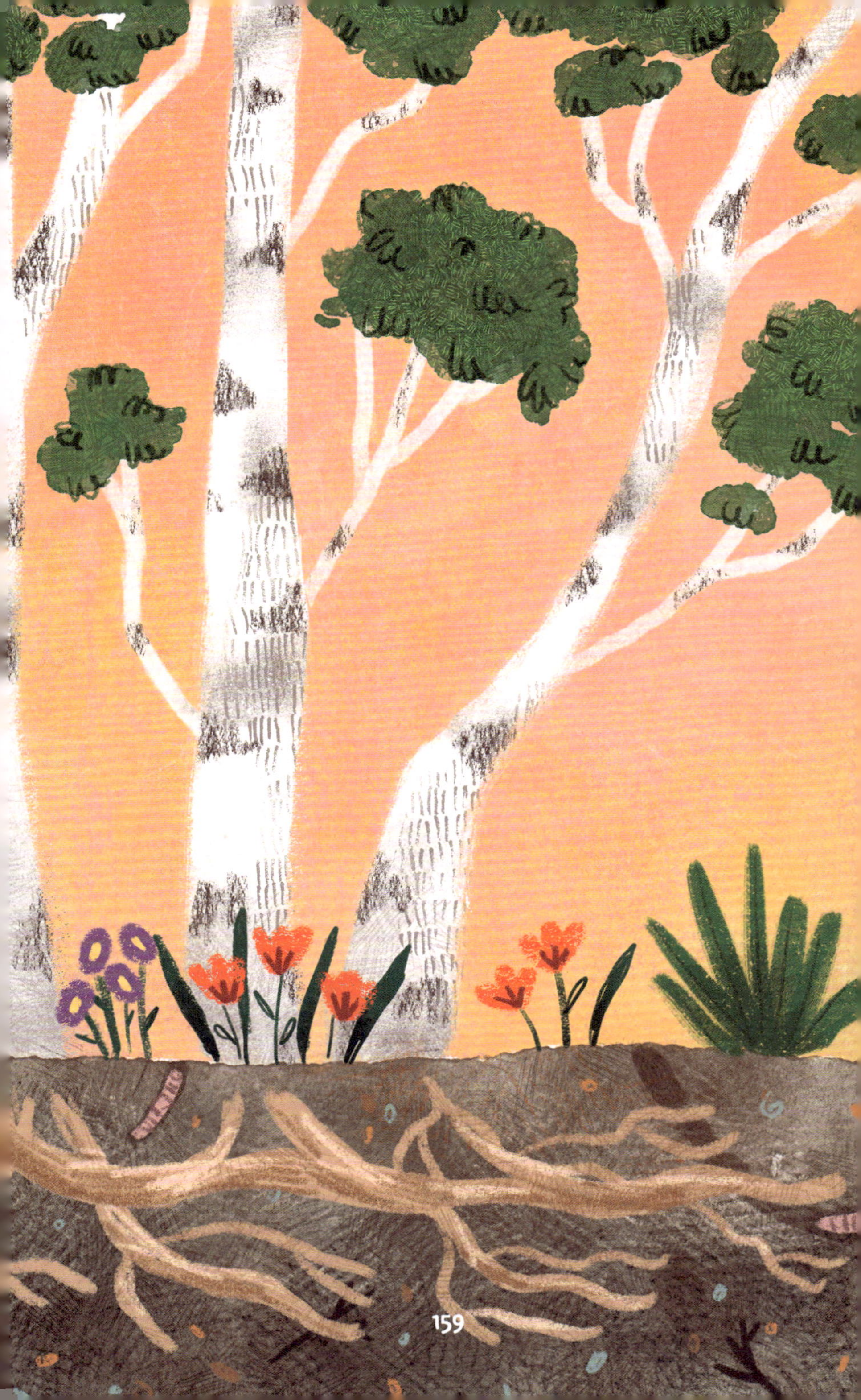

Troy admires the garden he grew. He'll see the roots when he picks all the things at harvest time.

Roots take in the water and vitamins the plants need.

We couldn't enjoy these lovely plants without roots.

Book 18

GEMS

Soft C and G; Sight Words: Group 4

The Good and the Beautiful

Illustrated by Larissa Sharina
© 2022 The Good and the Beautiful, LLC
goodandbeautiful.com

For ages, people have been interested in gems. Gems are both rare and pleasing to the eye.

Gems are full of colors: red, blue, pink, yellow, orange, and more.

Azurite

Andradite

Barite

Amber

Gems are not always hard. Some of these soft gems need to be polished gently.

Gems can be found in places like icy Greenland and the center of hot Africa.

Some gems look strange and fancy.

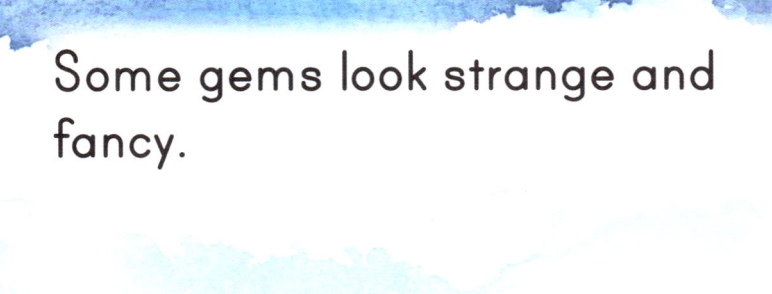

YOUNG GEOLOGIST 01/2022

MOST VALUABLE GEMSTONES

- DIAMOND
- RUBY
- SAPPHIRE
- EMERALD
- ALEXANDRITE

AFFORDABLE GEMSTONES

- SPINEL
- AQUAMARINE
- OPAL
- AMETHYST
- TURQUOISE
- PERIDOT
- AGATE

STONES

- AMBER
- LAPIS LAZULI
- JADE

The price for one gem can be very high, while the price for another gem is not high.

Book 19

THE SHETLAND PONY

OA and OE; OW Can Make the Long Sound O

The Good and the Beautiful

Good and the Beautiful

Illustrated by Shannon Vogus
© 2022 The Good and the Beautiful, LLC
goodandbeautiful.com

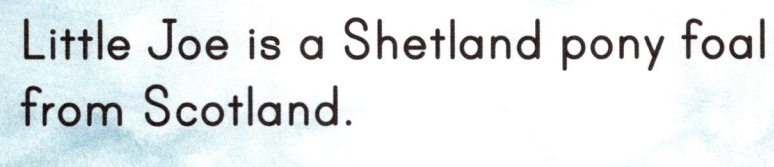

Little Joe is a Shetland pony foal from Scotland.

Look at some of the colors a Shetland pony can be.

They are small but so strong that people once used them to move big loads in coal mines.

As Joe grows, his coat becomes thick, keeping him warm in the winter.

In the summer he sheds much of his coat.

He eats hay and oats, but he loves to go up the road and roam the hillside and eat grass.

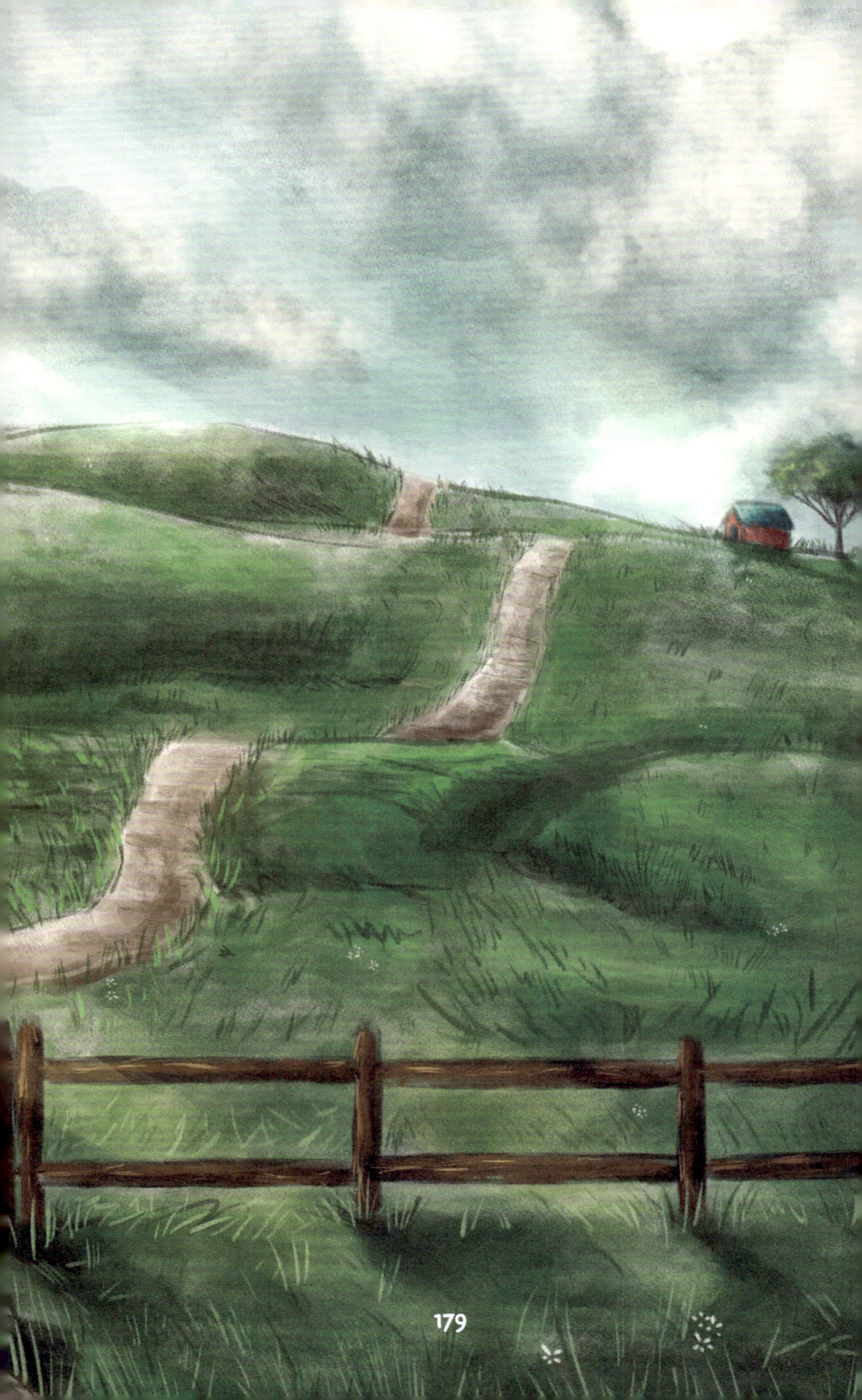

Joe does not have toes, but he has hooves. Look at some other parts of a Shetland pony.

ears
mane
withers
flank
tail
belly
muzzle
hoof

BOOK 20

THE WEATHER

Other Sounds of EA;
Words with Open Syllables

The Good and the Beautiful

Illustrated by Shannon Vogus
© 2022 The Good and the Beautiful, LLC
goodandbeautiful.com

Weather can be as soft as a feather, with the breeze fanning the flowers in the meadow.

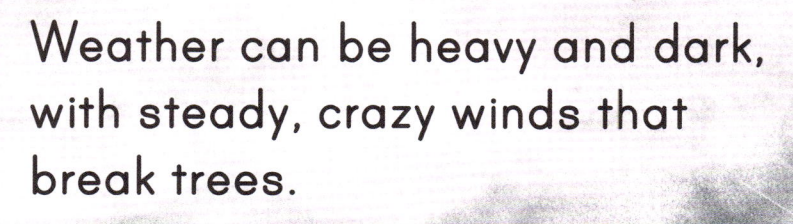

Weather can be heavy and dark, with steady, crazy winds that break trees.

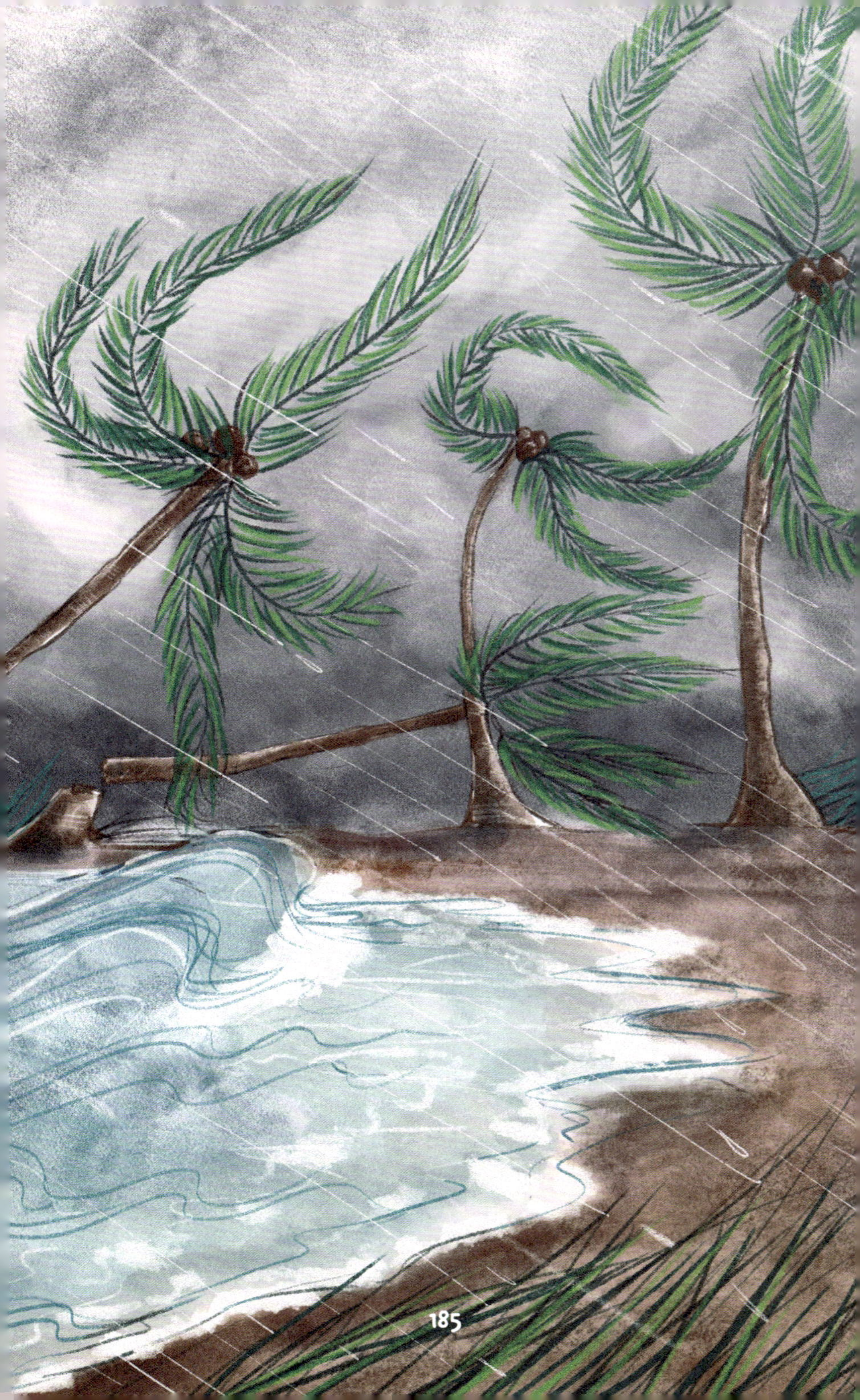

When weather is calm, it is silent here. On other days clouds spread across the sky.

The rain patters on the ground like music.

Weather can send its icy breath across a frozen land.